LET'S FIND OUT! OUR DYNAMIC EARTH

# THE CARBON CYCLE



LAURA LORIA

Britannica Educational Publishing
IN ASSOCIATION WITH
ROSEN
EDUCATIONAL SERVICES

Published in 2018 by Britannica Educational Publishing (a trademark of Encyclopædia Britannica, Inc.) in association with The Rosen Publishing Group, Inc.
29 East 21st Street, New York, NY 10010



Distributed exclusively by Rosen Publishing.
To see additional Britannica Educational Publishing titles, go to rosenpublishing.com.

First Edition

**Britannica Educational Publishing**
J.E. Luebering: Executive Director, Core Editorial
Mary Rose McCudden: Editor, Britannica Student Encyclopedia

**Rosen Publishing**
Amelie von Zumbusch: Editor
Nelson Sá: Art Director
Nicole Russo-Duca: Designer & Book Layout
Cindy Reiman: Photography Manager
Karen Huang: Photo Researcher

**Library of Congress Cataloging-in-Publication Data**

Names: Loria, Laura, author.
Title: The carbon cycle / Laura Loria.
Description: New York: Britannica Educational Publishing, in Association with Rosen Educational Services, 2018. | Series: Let's find out! Our dynamic earth | Audience: Grades 1–4. | Includes bibliographical references and index.
Identifiers: LCCN 2017014536| ISBN 9781680488210 (library bound : alk. paper) | ISBN 9781680488203 (pbk. : alk. paper) | ISBN 9781538300251 (6 pack : alk. paper)
Subjects: LCSH: Carbon cycle (Biogeochemistry)—Juvenile literature.
Classification: LCC QH344 .L67 2018 | DDC 577/.144—dc23
LC record available at https://lccn.loc.gov/2017014536

*Manufactured in the United States of America*

**Photo credits**: Cover Designua/Shutterstock.com; p. 4 Mariam Doerr Martin Frommherz/Shutterstock.com; p. 5 kali9/E+ /Getty Images; p. 6 tntphototravis/Shutterstock.com; p. 7 Inti St Clair/Blend Images/Getty Images; p. 8 © Courtesy, Image Science & Analysis Laboratory, NASA Johnson Space Center, No. ISS001-421-24; pp. 9, 14, 21, 27 © Encyclopædia Britannica, Inc.; p. 10 Jupiterimages/PHOTOS.com/Thinkstock; p. 11 igoraul/Fotolia; p. 12 Dziewul/Shutterstock.com; p. 13 © Tinichan/Fotolia; p. 15 Brazil Photos/LightRocket/Getty Images; p. 16 Jiri Foltyn/Shutterstock.com; p. 17 © David H. Harlow/U.S. Geological Survey; p. 18 Catherine Delahaye/The Image Bank/Getty Images; p. 19 Alice Fox/iStock/Thinkstock; p. 20 Xico Putini/Fotolia; p. 22 Science Source; p. 23 Michal Kowalski/Shutterstock.com; p. 24 © David Burdick/NOAA; p. 25 Darla Hallmark/Shutterstock.com; p. 26 Jim West/Science Source; p. 28 Vyacheslav Svetlichnyy/Shutterstock.com; p. 29 SkyLynx/Shutterstock.com; back cover and interior pages background © iStockphoto.com/ ithinksky.

# Contents


Life as We Know It 4
Carbon Reservoirs 6
High in the Sky 8
The Ocean Blue 10
Rock Solid 12
The Slow Carbon Cycle 14
The Fast Carbon Cycle 18
Changing the Atmosphere 20
Effects on Land 22
Effects on Oceans 24
The Study of Climate Change 26
Reducing Our Impact 28
Glossary 30
For More Information 31
Index 32

# Life as We Know It

Carbon is one of the most important chemical elements. Chemical elements are the building blocks for all matter. They are basic substances that cannot be broken down into simpler substances. Scientists use symbols to stand for the chemical elements. The symbol for carbon is C.

On its own, carbon makes up only a small part of Earth's crust. But it combines with other elements very easily to form substances called compounds. Carbon compounds make up 18 percent of all the matter in living things.

**Graphite is one of the forms of pure carbon that can be found in nature. It is used to make the leads of pencils.**

Living things are all around us. You are a living thing. So are animals and plants. Living things move, grow, and take in chemicals and nutrients.

**VOCABULARY**

A **cycle** is a series of events or actions that repeat regularly.

Carbon is constantly being used and replaced in nature through what is known as the carbon cycle. Carbon passes through living things, as well as through Earth's rocks, water, and air. Sometimes it moves very quickly, while other times it can stay in one place for many years.

All living things, including people and other animals, are part of the carbon cycle.

# Carbon Reservoirs

Earth has many carbon reservoirs, or places where carbon exists in large amounts. Carbon is found in Earth's hydrosphere (waters), atmosphere (the layer of gases surrounding the planet), lithosphere (rocky outer layer), and biosphere (where life happens).

Without carbon, life as we know it would not exist. All plants and animals are made of tissues. These tissues are built of elements grouped around chains

**This northern flicker's tissues contain carbon. So do the tissues of the saguaro cactus in which the bird lives.**

**Humans and plants exchange carbon dioxide. People breathe it out, while plants absorb it.**

or rings made of carbon atoms.

Living things produce carbon dioxide. People and other animals breathe out the carbon dioxide they make. Plants take in more carbon dioxide than they produce. They use it in photosynthesis, the process in which they make their own food. Photosynthesis requires sunlight, chlorophyll (a substance in plants and certain algae), water, and carbon dioxide. It produces oxygen and carbon compounds called sugars, which plants use and store as food. People and other animals breathe in oxygen and depend on the sugars plants make for food.

**COMPARE AND CONTRAST**

**How are animals and plants similar? How are they different?**

# High in the Sky

Earth's atmosphere consists of several gases. It has five layers. Near Earth's surface, the atmosphere is about three-fourths nitrogen and one-fifth oxygen. At higher elevations, the atmosphere is mostly hydrogen and helium.

**Carbon dioxide is present in each of the different layers of the atmosphere, but there is more of it near Earth's surface.**

The atmosphere does not contain pure carbon, but it does have carbon dioxide and methane. Methane is another compound that contains carbon. Carbon dioxide and methane enter the atmosphere by things people do and by natural processes.

Carbon dioxide and methane are both greenhouse gases. They are called this because they keep heat in a certain area, like a greenhouse does. Land, oceans, and plants absorb, or soak up, energy from sunlight. They release some of this energy as heat. Greenhouse gases absorb the heat and then send it back toward Earth. This is called the greenhouse effect. Without greenhouse gases, this heat would escape back into space, and Earth would be too cold for life to exist.

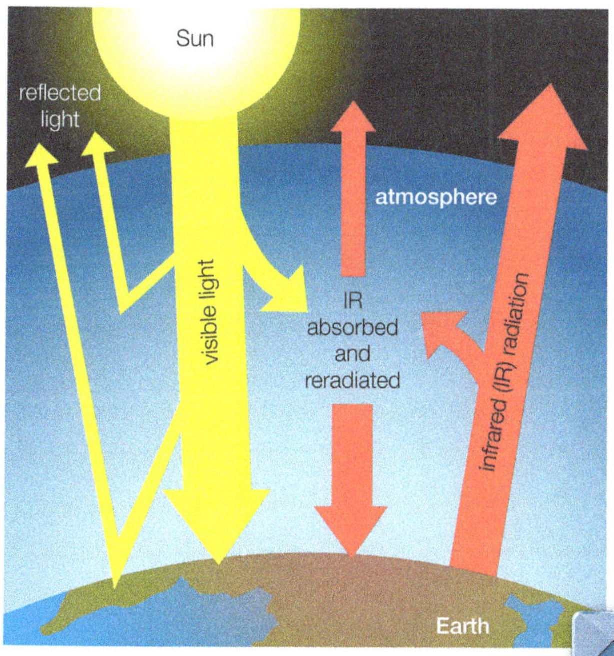


## Think About It

**What is the original source of the heat that greenhouse gases trap?**

**Heat is trapped and sent back to Earth by greenhouse gases, like carbon dioxide.**

# The Ocean Blue

The oceans and other bodies of water on Earth make up the hydrosphere. The oceans are important carbon reservoirs. They keep much more carbon than they give out. In the carbon cycle, carbon has a long stay in the oceans.

Carbon is found in several forms in an ocean. Carbon dioxide from the atmosphere dissolves into the surface waters. Tiny ocean

**The ocean's surface waters hold less carbon dioxide than its deeper waters.**

organisms called phytoplankton use the carbon dioxide in the water for photosynthesis. Other sea creatures eat phytoplankton. Their bodies combine carbon from the phytoplankton with calcium to make calcium carbonate. This white, powdery compound makes up the shells and skeletons of many ocean animals.

Most of the carbon in the ocean is found in deep water. When a sea creature dies, its body sinks to the bottom. The body decays, or breaks down, and carbon from the body tissue is released. The bits of carbon, with other materials, form layers that become rock over a long period of time.

**VOCABULARY**

**Organisms** are living things. People, plants, fungi, and algae are all organisms.

**The skeletons of stony corals are almost pure calcium carbonate.**

# Rock Solid

In the lithosphere, carbon is found in rocks and minerals. Rocks do not use the carbon like living things do. Carbon is stored in the rock. The two main forms of pure carbon in rocks are the minerals diamond and graphite. However, most of the carbon in rocks is in the form of carbon compounds.

Fossil fuels, which are buried in the lithosphere, are another carbon reservoir. Fossil fuels are the remains

These hikers are climbing over huge deposits of carbon in rock. About 99.9 percent of all near-surface carbon is stored in Earth's rocks.

## Compare and Contrast

**Coal is both a rock and a fossil fuel. How is it like other fossil fuels? How is it different?**

of organisms that lived long ago. When plants and animals die, their remains slowly decay. Eventually, they become buried under layers of soil. Over millions of years, they change to form different kinds of fossil fuels. Coal comes from plants. Petroleum and natural gas come mainly from microscopic organisms such as algae.

**Diamonds are crystals of pure carbon. They are formed deep within Earth under great amounts of pressure and heat.**

# The Slow Carbon Cycle

The carbon that is stored in rock is part of the geological carbon cycle. This is also called the slow carbon cycle, because it can take hundreds of millions of years for carbon to complete the cycle.

In the atmosphere, carbon combines with water vapor to form a kind of acid. The acid falls to Earth in rain. On the

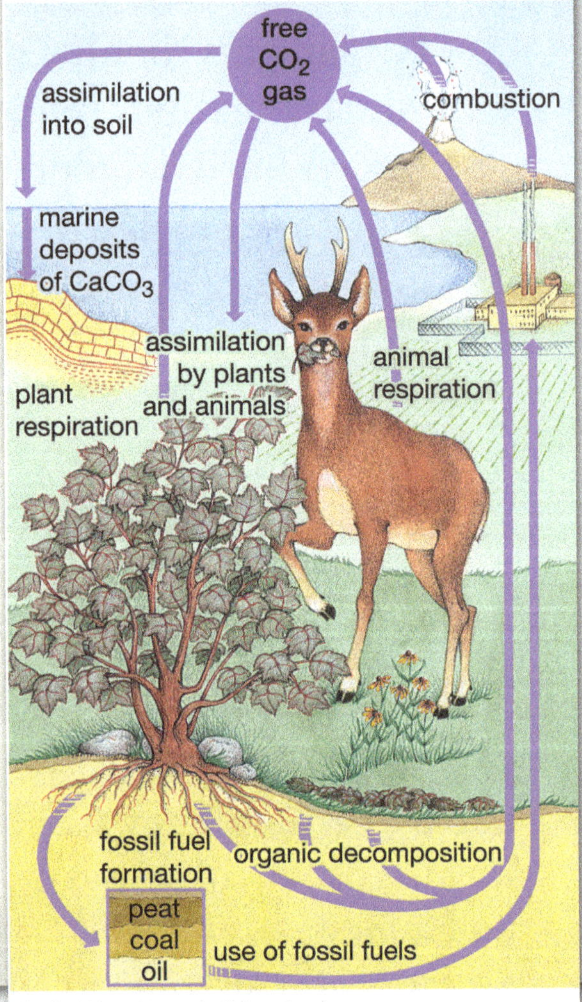

© 2014 Encyclopædia Britannica, Inc.

**Carbon continues to move in its different forms through the atmosphere, oceans, and soil; through living things; and through fossil fuel formation and use.**

Calcium carbonate dripped and hardened into rock, creating this pattern.

ground, the acid dissolves rock very slowly. This process, called chemical weathering, releases elements, like calcium, from the rock. These elements are carried to the ocean by running water.

The bodies of some sea animals combine the calcium in the ocean with the carbon they get from eating plants. That forms calcium carbonate that forms their shells. When the animals die, their bodies sink to the bottom. As they layer and decay, a type of rock called sedimentary rock is formed. The carbon is trapped in it.

**VOCABULARY**

**Sedimentary rock** is produced when tiny bits, called sediment, are pressed together to make rock.

**The Pancake Rocks, in New Zealand, are made of limestone—a type of rock that forms from the calcium carbonate of shells and corals on the ocean floor.**

Scientists think that the floor of the ocean spreads over time, pushing rock at the bottom of the ocean toward dry land. This process takes millions of years.

The crust of Earth is not one solid layer. It is broken up into many pieces, called plates. Plates meet in many different locations on Earth, on land and in the ocean. Because the plates are constantly moving, rock is pushed under other layers of rock all the time.

Pressure builds when plates move this way, and the result is a volcanic eruption. A volcano releases lava,

## Think About It

**How are the plates of Earth's crust related to volcanic eruptions?**

or hot liquid rock, that cools and hardens on Earth's surface. It also releases millions of tons of carbon dioxide back into the atmosphere. From there, the cycle begins again.

Carbon dioxide and ash rise from Mount Pinatubo, a volcano in the Philippines, in 1991.

# The Fast Carbon Cycle

Carbon moves around the biosphere much more quickly than it moves through the geological carbon cycle. This is why the biological carbon cycle is also called the fast carbon cycle.

Living things get the energy they need from a process called respiration. This process happens in the cells, so it is also called cellular respiration. Most living things—including plants and animals—need oxygen for respiration. They also need a simple sugar,

**Every time you breathe in or out, you are taking part in the fast carbon cycle.**

Photosynthesis happens mainly in the green leaves of plants.

called glucose. Plants (and some kinds of algae) make oxygen and glucose during photosynthesis. Animals breathe in oxygen from the air and get glucose from the foods they eat.

**COMPARE AND CONTRAST**

**How are the geological carbon cycle and the biological carbon cycle similar? How are they different?**

Cellular respiration takes place when cells take in oxygen and sugar and release water and carbon dioxide. During this process, energy is also released.

Animals breathe out the carbon dioxide that was released from their cells. Plants take in carbon dioxide from the air to use in photosynthesis. The process of photosynthesis produces glucose, which is used by animals. And the biological carbon cycle continues.

# Changing the Atmosphere

Human beings are part of the carbon cycle when they breathe out carbon dioxide. They also affect the cycle through other actions.

When people burn fossil fuels, the process releases carbon that had been stored deep inside Earth. Cutting down trees (for wood or to make room for homes and businesses) means there are fewer trees to take in carbon dioxide. This leads to more carbon dioxide in the atmosphere.

When there are fewer plants to absorb carbon dioxide, more of the gas will stay in the atmosphere.

More carbon dioxide in the atmosphere means that more heat gets trapped near Earth. Without human activity, Earth does a good job of balancing the amount of carbon that moves through the cycle. Since humans started adding carbon dioxide to the atmosphere, Earth's surface temperatures have begun to slowly rise. This affects all living things on our planet.

**THINK ABOUT IT**

**Do you think farms or forests would absorb more carbon dioxide? Why?**

**Human-Induced Carbon Dioxide Emissions**

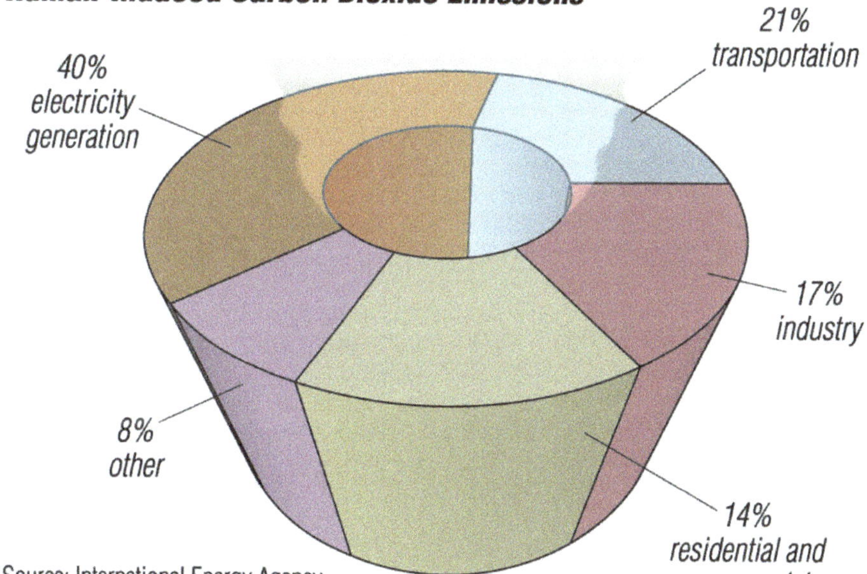


Source: International Energy Agency

**As this graph shows, there are many human-made sources of carbon dioxide.**

# Effects on Land

These scientists are testing plants to see the effects of carbon fertilization.

As the planet's surface gets warmer, plants adjust how they grow. Warmer temperatures mean a longer growing season. That means that more food can be produced for humans and other animals. Scientists have discovered that plants are actually absorbing more carbon dioxide than they have in the past. This is called carbon fertilization.

It sounds like more plant growth is a good

## Think About It

**Why does more carbon dioxide in the atmosphere only cause a temporary increase in plant growth?**

thing, but there are some problems. Plants don't just use carbon dioxide to grow. They also need water and nutrients like nitrogen from the soil. When a plant uses more carbon dioxide, it also uses more water and nutrients. This can dry out the soil in which the plant grows. Dry soil with few nutrients does not lead to more plants growing in the future.

**Dried-out soil makes it hard for plants to grow. So does depleted soil, or soil that has few nutrients left in it.**

# Effects on Oceans

The hydrosphere is home to many organisms. They are affected by any changes in the water. The extra heat trapped near Earth by extra carbon dioxide in the atmosphere makes for warmer waters, too. Warmer oceans increase the growth of some organisms and make it harder for others to survive.

Extra carbon dioxide dissolved in the water makes the ocean more acidic. One effect of this is that animals have a harder time making their shells. Shells

Coral bleaching occurs when oceans warm. The corals' main food, algae, leaves. Weak from little food, the corals turn white.

## Think About It

**Hotter temperatures also make Earth's oceans saltier. What effects could that have?**

are a type of protection for these animals, so they are more likely to die if they have weak shells.

A rise in air and water temperatures also causes glaciers and sea ice to melt, which makes the water level of the ocean rise. As the water level rises, more land on the shores is covered with water. This could mean that people who live on islands and coasts would lose their homes.

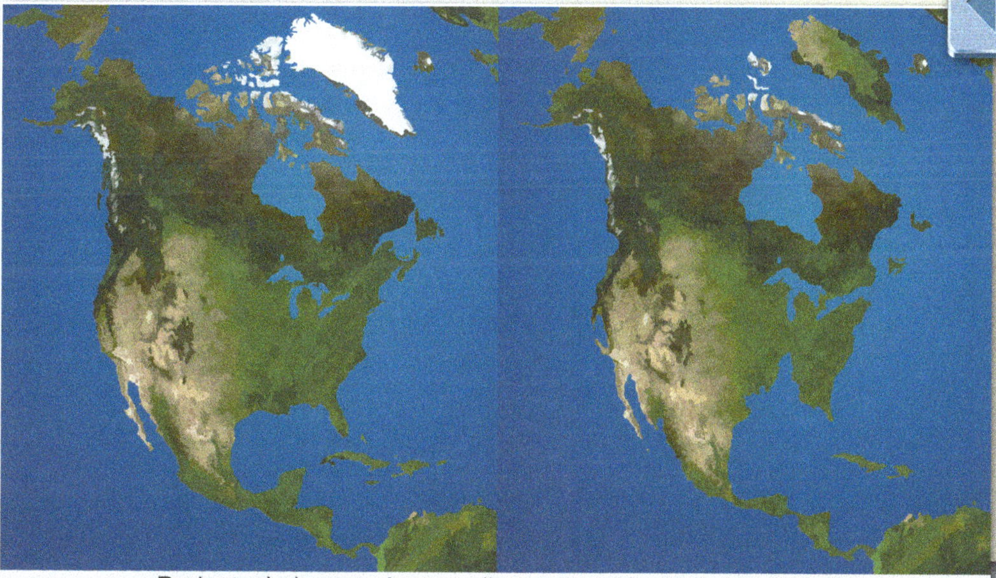

Projected changes in coastline, caused by rising sea levels.

**These images show North America today (*left*) and what it might look like if ocean levels rise (*right*).**

# The Study of Climate Change

Scientists study how extra carbon dioxide in the atmosphere affects our planet. The gradual warming of Earth's air and waters is called climate change. To understand how our planet's climate has changed, the scientists must first search for clues about what the climate was like in the past.

They can look at layers of rock for clues. The oldest layers are deep inside Earth. Scientists can also study gas bubbles in glaciers and the rings found in older trees. They also look for patterns by studying weather

This scientist is using a device to measure levels of carbon dioxide near the ground.

This graph shows rising carbon dioxide levels recorded at Hawaii's Mauna Loa Observatory.

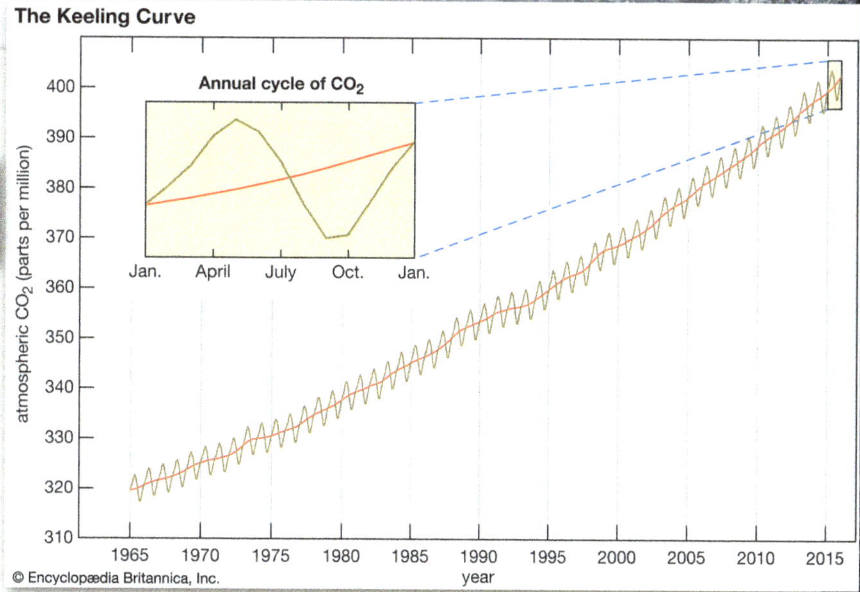


information that has been gathered over time.

Scientists have found that, throughout Earth's past, temperatures have gone up and down at different times. Some people think that this proves that human activity does not have much of an effect on climate. However, scientists say that the climate is changing much more quickly than it did in the past. This is likely to continue in the future.

### Think About It

What are some of the effects of warmer temperatures on Earth?

# Reducing Our Impact

People use coal, oil, and natural gas to run power plants, machines in factories, and cars. Fossil fuels make some things in our lives easier. However, burning fossil fuels can also be harmful to Earth. This can make our lives more difficult.

Reducing our use of fossil fuels is a good idea for several reasons. When less fossil fuels are burned, less carbon dioxide is released into the air. Also,

**Coal is a fossil fuel. The coal people mine and use today began forming about 300 million years ago.**

removing fossil fuels from the ground causes pollution and can be dangerous. Finally, the fossil fuels buried in Earth are nonrenewable, which means that they can be used up.

**Vocabulary**

**Solar energy** is energy from the sun.

Using renewable sources of energy (or energy that does not run out) can help reduce the amount of carbon we take from the ground and release into the air. Solar energy and wind power are two types of renewable power sources.

Windmills capture moving energy from wind and transform it into electrical energy.

# GLOSSARY

**acidic** Having a chemical that can break down substances.

**algae** Plantlike organisms, such as seaweed, that grow in water.

**atmosphere** The layer of gases that surrounds Earth or another planet.

**atom** One of the tiny particles that are the building blocks of all matter.

**biosphere** Earth's living organisms and the environments in which they live.

**chemical** A basic substance that reacts with other substances in a predictable way.

**compound** Something made up of two or more elements.

**decay** To rot, decompose, or break down.

**dissolve** To evenly mix with one or more other substances.

**element** A natural substance that cannot be broken down into other substances.

**glacier** A large body of ice on land.

**greenhouse** A building used for growing plants and featuring a glass roof.

**hydrosphere** All of the water at or near Earth's surface.

**lithosphere** The hard, rocky outside of Earth.

**microscopic** Too small to be seen without a microscope.

**mineral** A chemical element or compound that results from natural, nonliving processes. Rocks are made of minerals.

**nutrient** A substance that helps living things grow.

**oxygen** A gas found in air that is needed for life.

**photosynthesis** The process by which green plants use sunlight to make their own food.

**phytoplankton** Tiny organisms that live in water and can make their own food using photosynthesis. Phytoplankton are mostly algae.

**reservoir** A place where something is stored or exists in large amounts.

**respiration** The process that takes place in the cells of all living things that creates the energy needed to live.

**tissue** A group of cells that work together to perform a function.

# For More Information

## Books

Belton, Blair. *How Coal Is Formed.* New York, NY: Gareth Stevens Publishing, 2016.

Bow, James. *Earth's Climate Change: Carbon Dioxide Overload.* St. Catharines, ON, Canada: Crabtree Publishing Company, 2015.

Dakers, Diane. *The Carbon Cycle.* St. Catharines, ON, Canada: Crabtree Publishing Company, 2014.

Grosso, Valerie. *The Adventures of Carbon.* New York, NY: Be Naturally Curious, 2015.

Lindeen, Mary. *Investigating the Carbon Cycle.* Minneapolis, MN: Lerner Publications, 2015.

## Websites

Because of the changing nature of internet links, Rosen Publishing has developed an online list of websites related to the subject of this book. This site is updated regularly. Please use this link to access the list:

http://www.rosenlinks.com/LFO/Carbon

# Index


atmosphere, 6, 8–9, 18, 20–21

biological (fast) carbon cycle, 18–19
biosphere, 6, 18

calcium carbonate, 11, 15
carbon, chemical symbol for, 4
carbon compounds, 4, 12
carbon cycle, 5, 10, 14, 18-19, 20
carbon dioxide, 7, 8–9, 10–11, 19, 20–21, 22-23, 24, 28
carbon fertilization, 22
carbon reservoirs, 6, 10, 12
chlorophyll, 7
climate change, study of, 26–27

fossil fuels, 12–13, 20, 28
  reducing use of, 28–29

geological (slow) carbon cycle, 14–17
greenhouse gases, 9

hydrosphere, 6, 10–11, 24–25

lithosphere, 6, 12–13

methane, 8–9

oceans, 10–11, 16, 24–25
oxygen, 7, 8, 18

photosynthesis, 7, 11, 19
phytoplankton, 11
plates, 16

renewable energy, 29
respiration, 18–19
rocks, 11, 12, 14–15, 16–17

sedimentary rock, 15
tissues, 6–7

volcanoes, 16–17

warming of Earth's surface
  effects on oceans, 24–25
  effects on plants, 22–23
weathering, 14–15